YOUR KNOWLEDGE HAS VALUE

- We will publish your bachelor's and master's thesis, essays and papers

- Your own eBook and book - sold worldwide in all relevant shops

- Earn money with each sale

Upload your text at www.GRIN.com
and publish for free

Tushar Chatterji

Role of Nutritional Supplements in promoting Muscle Hypertrophy

GRIN Verlag

Bibliografische Information der Deutschen Nationalbibliothek:

Die Deutsche Bibliothek verzeichnet diese Publikation in der Deutschen Nationalbibliografie; detaillierte bibliografische Daten sind im Internet über http://dnb.d-nb.de/ abrufbar.

Dieses Werk sowie alle darin enthaltenen einzelnen Beiträge und Abbildungen sind urheberrechtlich geschützt. Jede Verwertung, die nicht ausdrücklich vom Urheberrechtsschutz zugelassen ist, bedarf der vorherigen Zustimmung des Verlages. Das gilt insbesondere für Vervielfältigungen, Bearbeitungen, Übersetzungen, Mikroverfilmungen, Auswertungen durch Datenbanken und für die Einspeicherung und Verarbeitung in elektronische Systeme. Alle Rechte, auch die des auszugsweisen Nachdrucks, der fotomechanischen Wiedergabe (einschließlich Mikrokopie) sowie der Auswertung durch Datenbanken oder ähnliche Einrichtungen, vorbehalten.

Imprint:

Copyright © 2011 GRIN Verlag GmbH
Druck und Bindung: Books on Demand GmbH, Norderstedt Germany
ISBN: 978-3-656-14994-1

This book at GRIN:

http://www.grin.com/en/e-book/190134/role-of-nutritional-supplements-in-promoting-muscle-hypertrophy

GRIN - Your knowledge has value

Der GRIN Verlag publiziert seit 1998 wissenschaftliche Arbeiten von Studenten, Hochschullehrern und anderen Akademikern als eBook und gedrucktes Buch. Die Verlagswebsite www.grin.com ist die ideale Plattform zur Veröffentlichung von Hausarbeiten, Abschlussarbeiten, wissenschaftlichen Aufsätzen, Dissertationen und Fachbüchern.

Visit us on the internet:

http://www.grin.com/

http://www.facebook.com/grincom

http://www.twitter.com/grin_com

Role of nutritional supplements in promoting muscle hypertrophy

Tushar Chatterji

College of Medical, Veterinary and Life Sciences

University of Glasgow

April 2011

Introduction

The skeletal muscle is an integral part of our system. It not only acts as the storage reservoir of amino acids, but also serves as the site for protein synthesis and protein breakdown [36]. The rate of protein synthesis needs to exceed protein degradation to achieve muscle hypertrophy [18, 24]. The timing of protein intake, type and quantity play a significant role in achieving optimal outcomes when applied to resistance exercise [36]. Research has been going on since the past decade demonstrating the role of nutritional supplements like whey protein, soy, branched-chain amino acids (BCAAs, especially leucine) and creatine on protein synthesis before, during and after a bout of resistance exercise through careful investigations into intracellular signalling pathways like the mammalian target of rapamycin (mTOR) and its downstream targets-ribosomal protein S6 (kinase-1) and 4E binding protein (4E-BP1) [7, 18]. Intracellular signalling, amongst other variables, involves three essential components- abundant ATP in muscle for providing energy, insulin signalling and leucine (figure 1) [36, 37]. mTOR, regarded as the "key regulator" of translation comprises mTORC1 and mTORC2 [36]. mTORC1 plays a significant role in promoting muscular hypertrophy via phosphorylation of S6K1 and 4E-BP1 which prevent binding of the eukaryotic initiation factor (eIF) 4E to 4E-BP1 allowing a complex formation with eIF4G (eIF4E-eIF4G) thus enhancing protein synthesis [5, 34].

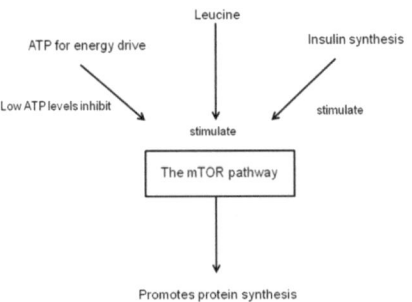

Figure 1. The three essential components for signalling: Leucine, insulin and ATP affect intracellular signalling of mTOR, thereby stimulating the rate of protein synthesis. Leucine directly and indirectly (via insulin synthesis) activates mTOR pathway while reducing ATP levels within the cell reduces the rate of protein synthesis [36].

Another important signalling pathway affected by nutrition and resistance exercise is the extracellular signal-regulated kinase ½ (ERK1/2) which stimulates translation independent of mTOR via mitogen activated protein kinase (MAPK) signalling to eIF4E (figure 2) [9, 35]. Insulin signalling indirectly phosphorylates the phosphatidylinositol-3-kinase (PI-3K) pathway facilitating glucose transport into the cell via a glucose transporter (GLUT4) [23]. Akt (component of mTOR pathway) activates mTOR and inactivates glycogen synthase kinase (GSK-3) stimulating eIF2B and mTOR thereby enhancing protein synthesis [19]. Despite this, high insulin levels fail to achieve significant protein synthesis in the absence of high levels of amino acids [3, 36]. Lastly, leucine by far has proved to be the most effective of amino acids in stimulating protein synthesis either directly through mTOR phosphorylation or indirectly by activating mTOR via the insulin pathway [26, 36]. This report highlights how whey proteins and BCAA (leucine) affect muscle protein synthesis via intracellular signalling pathways thereby contributing significant effects on muscular hypertrophy.

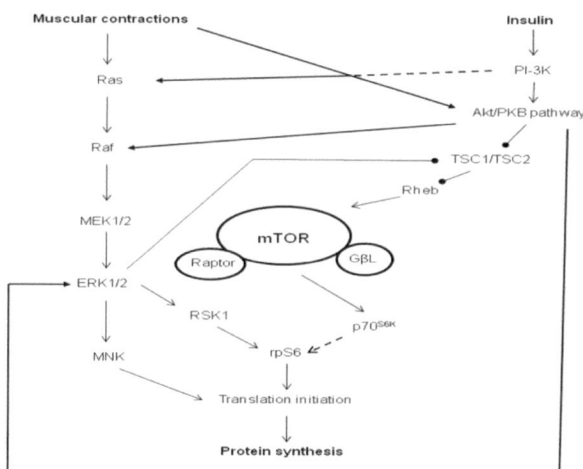

Figure 2. A diagrammatic representation of the interaction between mTOR and ERK1/2 in stimulating protein synthesis following muscular contractions and/or insulin synthesis: Contractions activate ERK1/2 via the Ras signalling cascade which eventually stimulates the formation of the translation initiation complex leading to protein synthesis. The figure also shows the components inhibited as a result of the signalling (arrows with rounded ends). The dashed arrow represents the possible conversion to rpS6. mTOR-mammalian target of rapamycin, MEK1/2-Mitogen-activated protein kinase1/2, ERK1/2-extracellular signal-regulated kinase1/2, MNK1-MAP kinase-interacting kinase1, PI-3K-phosphatidylinositol 3-kinase, PKB-protein kinase B, TSC1/TSC2-tuberous sclerosis complex, rpS6-ribosomal protein S6, Rheb-ras homolog enriched in brain [7, 16, 25].

The effects of whey protein supplementation

Whey protein, particularly whey protein concentrates (WPC 80) and isolates (WPI) have proved useful in individuals who are regularly training [8, 12]. Whey provides a quick and efficient recovery thereby enhancing performance, anaerobic fitness, muscle mass and body composition [12]. When applied to resistance exercise, whey protein supplementation has been found to increase muscle bulk by stimulating protein synthesis [8, 12, 18]. In a study demonstrating the effects of whey on subjects undergoing one session of resistance exercise, a rise in the phosphorylation of $p70^{S6K}$ and 4E-BP1 was observed [12]. mTOR phosphorylation remained high not only from 1 hour to 48 hours post exercise, but also after 21 weeks of resistance training, showing that whey plays a critical role in mTOR signalling. This aspect was further investigated in a double-blind, placebo controlled trial which looked into the effect of 20 grams of WPI ingestion on mTOR signalling in young men [8]. The results of this study were in conjunction with the former trial above demonstrating an increase in mTOR phosphorylation 2 hours post exercise following WPI intake. Activation of mTOR was brought about by the phosphorylation of Ser^{2448} via leucine and Akt [14, 15]. $p70^{S6K}$ was phosphorylated early on whey administration [8]. The increase in $p70^{S6K}$ phosphorylation by whey protein activated the TORC1 complex along with mTOR and its regulatory proteins [13, 35]. A probable reason for this mechanism might lie in the rich source of BCAA which make up the whey protein. The BCAA's evoke a similar response with respect to $p70^{S6K}$ after a bout of resistance exercise [4, 13]. Moreover, activation of $p70^{S6K}$ relies on the ability of these amino acids to stimulate the downstream signalling of rpS6 [14]. Higher doses of WPI would be required to activate rpS6 post exercise (i.e. during recovery). 4E-BP1 phosphorylation at $Thr^{37/46}$ (site-specific phosphorylation) increased by the ingestion of whey alone after 2 hours of resistance training [8, 12]. This prevention further dissociated 4E-BP1 from eIF4E, thus increasing protein synthesis [12]. These results suggested that 4E-BP1 phosphorylation could occur without depending on TORC1 complex unlike $p70^{S6K}$, reflecting that whey supplementation demonstrated TORC1 dependent and independent effects on mTOR signalling [13, 34, 35].

Signal transduction pathways such as PI3K/Akt activated by resistance training are responsible for eliciting skeletal hypertrophy (13,16). Akt is highly phosphorylated at Ser^{473} post 1 hour of resistance exercise for a short period of time (19,20). However, the two studies showed contradictory results- Farnfield *et al.* showed that Akt phosphorylation remained unchanged post resistance exercise on ingesting 26.6 grams WPI while Hulmi *et al.* reported a decline in Akt phosphorylation after 21 weeks of resistance training [8, 12]. This might be attributed to the differences in the duration of exercise, type of training and nutritional status [12]. Whey protein intake was successful in maintaining sustained mTOR signalling in response to resistance exercise and training. Future studies should be aimed at defining the appropriate timings and dose regimens of WPI to stimulate protein synthesis, in relation to resistance exercise [8].

The role of BCAAs and leucine supplementation

Leucine, a branched-chain amino acid, has demonstrated anabolic actions by stimulating protein synthesis and reducing protein degradation not only in resting skeletal muscle, but also during resistance and endurance exercises [4, 13, 16, 25, 26]. Besides an important component of whey (14% leucine and 26% BCAA), intracellular levels of leucine promote translation initiation to stimulate protein synthesis, insulin signalling via PI3K pathway and production of non essential amino acids alanine and glutamine in muscle [26, 37]. The role of leucine on muscle protein synthesis is independent of the effects of the enzyme branched-chain aminotransferase (BCAT) present in liver [11]. This allows leucine to travel through the bloodstream unperturbed depending on the dietary intake [1]. An increased flow of leucine from the liver to the skeletal muscle during resistance training accounted for the continuous influx of leucine to muscle.

Administration of leucine in human skeletal muscle at rest not only increased the phosphorylation of p70S6K and 4E-BP1, but also left the Akt and GSK-3 pathways inactivated for 2-6 hours [4, 10]. This result was confirmed by infusing a BCAA mixture after 6 hours [21]. When related to resistance training, a reduction in Akt phosphorylation was observed post exercise while mTOR and GSK-3 phosphorylation remained unchanged. The phosphorylation of Thr^{389}, which was also unchanged post exercise increased when BCAA combined with leucine after 1-2 hours of resistance exercise. BCAA ingestion activated mTOR on Ser^{2448} 1 hour post exercise without any significant effect on Akt and/or GSK-3 suggesting that Akt/GSK-3 pathway fails to induce an anabolic effect on human skeletal muscle in the presence of BCAA [10, 21]. There is evidence linking the role of MAPK with increasing muscle protein synthesis via MAPK-integrating kinase-1 [4, 13]. An increase in the phosphorylation of ERK1/2, p38 MAPK and $p70^{S6K}$ (at Ser^{424}/Thr^{421}) was observed immediately after resistance exercise [4, 16]. However, baseline values were identical to ERK1/2 and p38 MAPK phosphorylation after 1 hour of exercise suggesting their effects to be temporary [12]. MAPK signalling seemed to have divergent effects because of differences in exercise training as highly trained subjects would experience greater signalling responses to exercise [12, 38]. In other words, resistance exercise had a transient effect on MAPK signalling cascade. However, BCAA did not seem to have any effect on MAPK after resistance exercise suggesting that the phosphorylation of p70^{S6K} after BCAA ingestion was independent of MAPK signalling [12].

Leucine supplementation is believed to promote protein synthesis via an mTOR independent pathway too [26]. This was brought about by increased eIF4G phosphorylation which raised the affinity of eIF4G for eIF4E, facilitated eIF4G-eIF4E complex formation and stimulated protein synthesis independent of mTOR [3]. These findings suggested the importance of intracellular leucine concentrations not only for stimulating eIF4G, but also for providing sufficient eIF4E for maximal protein synthesis [26]. Activation of the mTOR pathway by leucine is influenced by the following proteins-tuberous sclerosis complex (TSC1/TSC2), ras-homolog in brain (Rheb), AMP kinase (AMPK) and raptor (mTOR regulatory protein) [16, 22, 26]. Rheb, a GTPase, increases phosphorylation of 4E-BP1 and S6K1 via mTOR [5].

A recent study demonstrated that Rheb formed a complex (Rheb-mTOR) by directly binding to it and its interacting protein (GβL/mLST8) without the need for GDP or GTP [22]. However, lack of GDP or GTP reduced mTOR signalling suggesting their importance for Rheb function in vivo [16]. GTP when bound to Rheb increased mTOR phosphorylation to a greater extent than when Rheb associated with GDP. Rheb in association with TSC1/TSC2 is required for appropriate mTOR signalling via leucine/amino acids [16]. The TSC1/TSC2 complex also known as tuberin-hamartin complex allows conversion from Rheb-GTP to Rheb-GDP, thus impairing mTOR signalling [26]. This allows leucine to target sites downstream of TSC2, most probably through Rheb [22, 26]. The activity of TSC2 (tuberin) is controlled by the phosphorylation of PKB/Akt, p90^{rsk}, ERK and AMPK [16]. mTOR signalling is increased when the GTPase activity of TSC2 is inhibited via ERK, p90^{rsk} or PKB/Akt phosphorylation thus enhancing association of GTP to Rheb. On the other hand, activation by AMPK amplifies tuberin reducing mTOR signalling [22, 29, 34]. Thus, mammalian cells deprived of amino acids and lacking either TSC1 or TSC2 would facilitate S6K1 phosphorylation [29].

Changes in the amino acid concentrations affect the association of mTOR with raptor [16]. When leucine is supplemented to leucine-deficient cells, an enhanced association of raptor to mTOR occurs while addition of dithiobis propionate (cross-linking agent) to the same cells does not provide a similar effect, suggesting that leucine induces a conformational shift in mTOR-raptor complex from being stable and inactive to unstable and active [17]. This can also be attributed to the suppression of mTOR signalling through an impaired association of raptor and mTOR by rapamycin [16]. This might occur by blocking of a TOR signalling motif (TOS) on raptor by mTOR in leucine-deficient cells [16, 17]. This induces a change in either raptor or mTOR allowing TOS to phosphorylate 4E-BP1 and S6K1 and associate them to the mTOR-raptor complex [16].

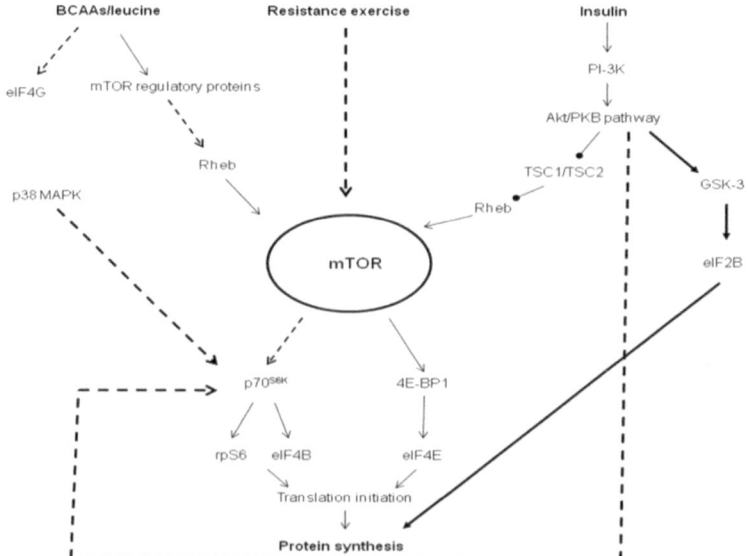

Figure 3. Schematic diagram demonstrating the effects of BCAA/leucine, resistance exercise and insulin on the mTOR signalling pathway leading up to protein synthesis: The figure

shows the way in which the mTOR pathway gets activated by leucine. Insulin plays a vital role during leucine supplementation. The dashed arrows represent the possible effects of components on mTOR which ultimately stimulate protein synthesis. TSC1/TSC2 and Rheb are inactivated in the process of mTOR activation (arrows with rounded ends). mTOR-mammalian target of rapamycin, eIF-eukaryotic initiation factor, PI-3K-phosphatidylinositol 3-kinase, TSC1/TSC2-tuberous sclerosis complex, Rheb-ras homolog enriched in brain, 4E-BP1-binding protein1, p38MAPK-p38 mitogen-activated protein kinase, GSK-3-glycogen synthase kinase3 [4, 7, 26].

Potential concerns with amino acid supplementation

It is interesting to note that amino acids when ingested undergo favourable oxidation, thereby suppressing glucose uptake by skeletal muscle despite stimulating higher insulin synthesis, a condition referred to as "glucose sparing" effect [25]. The association between the regulatory protein cofactor of PI-3K and tyrosine receptor of insulin substrates 1 and 2 not only inhibits insulin signalling and gluconeogenesis *in vitro*, but also reduces mTOR, PKB/Akt and PI-3K phosphorylation [27, 32]. Other concerns with BCAA supplementation are related to kidney functions and tumour progression involving mTOR and PI-3K pathways [25, 33]. Future research should target such areas and estimate potential boundaries for BCAA and/or leucine supplementation not only for promoting muscle hypertrophy, but also with respect to the health and safety of people.

Conclusion

In conclusion, considerable progress has been made in the area of health and fitness by understanding the cellular mechanisms behind nutritional supplementation in relation to exercise [7, 36]. Successful implementation of nutritional strategies such as leucine and whey supplementation in promoting protein synthesis post resistance training has significantly improved exercise performance. Efforts need to continue to maintain a synergistic association between protein ingestion and exercise to maximise protein synthesis and provide better muscular adaptations to exercise [36]. Issues concerning the upstream regulation of the mTOR pathway via amino acids and the corresponding salutary effects of BCAA and whey protein ingestion should be addressed in future studies [16, 25, 36].

References

1	Ahlborg, G., Felig, P., Hagenfeldt, L., Hendler, R. and Wahren, J. (1974) Substrate turnover during prolonged exercise in man. Splanchnic and leg metabolism of glucose, free fatty acids, and amino acids. J Clin Invest. 53, 1080-1090

2	Anthony, J. C., Lang, C. H., Crozier, S. J., Anthony, T. G., MacLean, D. A., Kimball, S. R. and Jefferson, L. S. (2002) Contribution of insulin to the translational control of protein synthesis in skeletal muscle by leucine. Am J Physiol Endocrinol Metab. 282, E1092-1101

3	Biolo, G., Declan Fleming, R. Y. and Wolfe, R. R. (1995) Physiologic hyperinsulinemia stimulates protein synthesis and enhances transport of selected amino acids in human skeletal muscle. J Clin Invest. 95, 811-819

4	Blomstrand, E., Eliasson, J., Karlsson, H. K. and Kohnke, R. (2006) Branched-chain amino acids activate key enzymes in protein synthesis after physical exercise. J Nutr. 136, 269S-273S

5	Bodine, S. C., Stitt, T. N., Gonzalez, M., Kline, W. O., Stover, G. L., Bauerlein, R., Zlotchenko, E., Scrimgeour, A., Lawrence, J. C., Glass, D. J. and Yancopoulos, G. D. (2001) Akt/mTOR pathway is a crucial regulator of skeletal muscle hypertrophy and can prevent muscle atrophy in vivo. Nat Cell Biol. 3, 1014-1019

6\tCastro, A. F., Rebhun, J. F., Clark, G. J. and Quilliam, L. A. (2003) Rheb binds tuberous sclerosis complex 2 (TSC2) and promotes S6 kinase activation in a rapamycin- and farnesylation-dependent manner. J Biol Chem. 278, 32493-32496

7\tDrummond, M. J., Dreyer, H. C., Fry, C. S., Glynn, E. L. and Rasmussen, B. B. (2009) Nutritional and contractile regulation of human skeletal muscle protein synthesis and mTORC1 signaling. J Appl Physiol. 106, 1374-1384

8\tFarnfield, M. M., Carey A. K., Gran Peter, Trennerry K. M., Smith-Cameron D. (2009) Whey Protein Ingestion Activates mTOR-dependent Signalling after Resistance Exercise in Young Men: A Double-Blinded Randomized Controlled Trial. Nutrients, 263-275

9\tFluckey, J. D., Knox, M., Smith, L., Dupont-Versteegden, E. E., Gaddy, D., Tesch, P. A. and Peterson, C. A. (2006) Insulin-facilitated increase of muscle protein synthesis after resistance exercise involves a MAP kinase pathway. Am J Physiol Endocrinol Metab. 290, E1205-1211

10\tGreiwe, J. S., Kwon, G., McDaniel, M. L. and Semenkovich, C. F. (2001) Leucine and insulin activate p70 S6 kinase through different pathways in human skeletal muscle. Am J Physiol Endocrinol Metab. 281, E466-471

11\tHarper, A. E., Miller, R. H. and Block, K. P. (1984) Branched-chain amino acid metabolism. Annu Rev Nutr. 4, 409-454

12\tHulmi, J. J., Tannerstedt, J., Selanne, H., Kainulainen, H., Kovanen, V. and Mero, A. A. (2009) Resistance exercise with whey protein ingestion affects mTOR signaling pathway and myostatin in men. J Appl Physiol. 106, 1720-1729

13\tKarlsson, H. K., Nilsson, P. A., Nilsson, J., Chibalin, A. V., Zierath, J. R. and Blomstrand, E. (2004) Branched-chain amino acids increase p70S6k phosphorylation in human skeletal muscle after resistance exercise. Am J Physiol Endocrinol Metab. 287, E1-7

14   Koopman, R., Zorenc, A. H., Gransier, R. J., Cameron-Smith, D. and van Loon, L. J. (2006) Increase in S6K1 phosphorylation in human skeletal muscle following resistance exercise occurs mainly in type II muscle fibers. Am J Physiol Endocrinol Metab. 290, E1245-1252

15   Koulmann, N. and Bigard, A. X. (2006) Interaction between signalling pathways involved in skeletal muscle responses to endurance exercise. Pflugers Arch. 452, 125-139

16   Kimball, S. R. and Jefferson, L. S. (2006) Signaling pathways and molecular mechanisms through which branched-chain amino acids mediate translational control of protein synthesis. J Nutr. 136, 227S-231S

17   Kim, D. H., Sarbassov, D. D., Ali, S. M., King, J. E., Latek, R. R., Erdjument-Bromage, H., Tempst, P. and Sabatini, D. M. (2002) mTOR interacts with raptor to form a nutrient-sensitive complex that signals to the cell growth machinery. Cell. 110, 163-175

18   Kumar, V., Atherton, P., Smith, K. and Rennie, M. J. (2009) Human muscle protein synthesis and breakdown during and after exercise. J Appl Physiol. 106, 2026-2039

19   Leger, B., Cartoni, R., Praz, M., Lamon, S., Deriaz, O., Crettenand, A., Gobelet, C., Rohmer, P., Konzelmann, M., Luthi, F. and Russell, A. P. (2006) Akt signalling through GSK-3beta, mTOR and Foxo1 is involved in human skeletal muscle hypertrophy and atrophy. J Physiol. 576, 923-933

20   Liu, Z., Jahn, L. A., Long, W., Fryburg, D. A., Wei, L. and Barrett, E. J. (2001) Branched chain amino acids activate messenger ribonucleic acid translation regulatory proteins in human skeletal muscle, and glucocorticoids blunt this action. J Clin Endocrinol Metab. 86, 2136-2143

21   Liu, Z., Wu, Y., Nicklas, E. W., Jahn, L. A., Price, W. J. and Barrett, E. J. (2004) Unlike insulin, amino acids stimulate p70S6K but not GSK-3 or glycogen synthase in human skeletal muscle. Am J Physiol Endocrinol Metab. 286, E523-528

22   Long, X., Lin, Y., Ortiz-Vega, S., Yonezawa, K. and Avruch, J. (2005) Rheb binds and regulates the mTOR kinase. Curr Biol. 15, 702-713

23   MacLean, P. S., Zheng, D. and Dohm, G. L. (2000) Muscle glucose transporter (GLUT 4) gene expression during exercise. Exerc Sport Sci Rev. 28, 148-152

24   Miyazaki, M. and Esser, K. A. (2009) Cellular mechanisms regulating protein synthesis and skeletal muscle hypertrophy in animals. J Appl Physiol. 106, 1367-1373

25   Nair, K. S. and Short, K. R. (2005) Hormonal and signaling role of branched-chain amino acids. J Nutr. 135, 1547S-1552S

26   Norton, L. E. and Layman, D. K. (2006) Leucine regulates translation initiation of protein synthesis in skeletal muscle after exercise. The Journal of nutrition. 136, 533S-537S

27   Patti, M. E., Brambilla, E., Luzi, L., Landaker, E. J. and Kahn, C. R. (1998) Bidirectional modulation of insulin action by amino acids. J Clin Invest. 101, 1519-1529

28   Raught, B. and Gingras, A. C. (1999) eIF4E activity is regulated at multiple levels. Int J Biochem Cell Biol. 31, 43-57

29   Saucedo, L. J., Gao, X., Chiarelli, D. A., Li, L., Pan, D. and Edgar, B. A. (2003) Rheb promotes cell growth as a component of the insulin/TOR signalling network. Nat Cell Biol. 5, 566-571

30   Tang, W., Yuan, J., Chen, X., Gu, X., Luo, K., Li, J., Wan, B., Wang, Y. and Yu, L. (2006) Identification of a novel human lysophosphatidic acid acyltransferase, LPAAT-theta, which activates mTOR pathway. J Biochem Mol Biol. 39, 626-635

31   Thomson, D. M., Fick, C. A. and Gordon, S. E. (2008) AMPK activation attenuates S6K1, 4E-BP1, and eEF2 signaling responses to high-frequency electrically stimulated skeletal muscle contractions. J Appl Physiol. 104, 625-632

32   Tremblay, F. and Marette, A. (2001) Amino acid and insulin signaling via the mTOR/p70 S6 kinase pathway. A negative feedback mechanism leading to insulin resistance in skeletal muscle cells. J Biol Chem. 276, 38052-38060

33   Vogt, P. K. (2001) PI 3-kinase, mTOR, protein synthesis and cancer. Trends Mol Med. 7, 482-484

34   Wang, X. and Proud, C. G. (2006) The mTOR pathway in the control of protein synthesis. Physiology (Bethesda). 21, 362-369

35   Wang, X., Yue, P., Chan, C. B., Ye, K., Ueda, T., Watanabe-Fukunaga, R., Fukunaga, R., Fu, H., Khuri, F. R. and Sun, S. Y. (2007) Inhibition of mammalian target of rapamycin induces phosphatidylinositol 3-kinase-dependent and Mnk-mediated eukaryotic translation initiation factor 4E phosphorylation. Mol Cell Biol. 27, 7405-7413

36   Weinert, D. J. (2009) Nutrition and muscle protein synthesis: a descriptive review. The Journal of the Canadian Chiropractic Association. 53, 186-193

37   Willoughby, D. S., Stout, J. R. and Wilborn, C. D. (2007) Effects of resistance training and protein plus amino acid supplementation on muscle anabolism, mass, and strength. Amino acids. 32, 467-477

38   Yu, M., Stepto, N. K., Chibalin, A. V., Fryer, L. G., Carling, D., Krook, A., Hawley, J. A. and Zierath, J. R. (2003) Metabolic and mitogenic signal transduction in human skeletal muscle after intense cycling exercise. J Physiol. 546, 327-335